Quick References for Warehouse Temperature and Humidity Control Validation

Quick Reference Series by makeiteasy901 Store

Contents

Step-by-Step Validation Plan 3

Example of Validation Plan Template 4

Step-by-Step Installation Qualification (IQ) 7

Installation Qualification (IQ) Template 8

Step-by-Step Operational Qualification (OQ) 11

Example Operational Qualification (OQ) Template 12

Step-by-Step Performance Qualification (PQ) 15

Example Performance Qualification (PQ) Template 16

Step-by-Step Risk Assessment 19

Example Risk Assessment Template...20

Step-by-Step Data Analysis...23

Example Data Analysis Template 24

Step-by-Step Determination for Revalidation 28

Example Revalidation Trigger Template 29

Step-by-Step Guide to Resolving Deviations During Validation....................32

Example: Resolving a Deviation During Validation.......................................33

Step-by-Step Non-Conformance Investigation and Correction 36

Example Non-Conformance Investigation and Correction Template..............37

Step-by-Step Validation Plan

1. Introduction

- **Objective:** Provide an overview of the validation plan, explaining the importance of maintaining controlled temperature and humidity in the warehouse.

- **Scope:** Define the areas within the warehouse that are subject to validation.

- **Regulatory Requirements:** Mention relevant regulatory guidelines and standards that the validation plan will adhere to.

2. Objectives

- State the main goals of the validation, such as ensuring product quality and compliance with regulatory requirements.

- Define specific objectives, like verifying that temperature and humidity are maintained within specified limits.

3. Scope

- Detail the specific zones, rooms, or sections of the warehouse included in the validation.

- Specify the equipment and systems (e.g., HVAC, sensors, data loggers) involved in maintaining and monitoring temperature and humidity.

4. Responsibilities

- List the roles and responsibilities of the personnel involved in the validation process.

- Include responsibilities for preparation, execution, review, and approval of validation activities.

5. Acceptance Criteria

- Define the acceptable ranges for temperature and humidity within the warehouse.

- Set criteria for pass/fail status for each validation activity (IQ, OQ, PQ).

6. Validation Activities

- **Installation Qualification (IQ):** Verify that all equipment is installed correctly and operates as per specifications.

  - List the equipment to be installed.

  - Describe the installation process and checks.

- **Operational Qualification (OQ):** Test the equipment to ensure it operates within the defined parameters under different conditions.

  - Detail the tests to be performed, including normal and worst-case scenarios.

- **Performance Qualification (PQ):** Validate that the warehouse maintains the required environmental conditions over a specific period.

  - Outline the monitoring plan and data collection methods.

7. Schedule

- Provide a timeline for each validation activity, from preparation to final approval.

- Include key milestones and deadlines.

8. Documentation

- List the documents to be prepared and maintained during the validation process, such as protocols, test results, and reports.

- Ensure that all documents are reviewed and approved by the responsible parties.

9. Approval

- Include a section for signatures of personnel responsible for preparing, reviewing, and approving the validation plan.

- Ensure that all necessary approvals are obtained before starting the validation activities.

Example of Validation Plan Template

Below is a template for the validation plan document.

Validation Plan for Temperature and Humidity Control in Warehouse

1. Introduction

- **Objective:** This validation plan outlines the process for validating temperature and humidity control in the warehouse to ensure compliance with regulatory standards and maintain product quality.

- **Scope:** This plan covers the validation of temperature and humidity control in [Specify areas] of the warehouse.

- **Regulatory Requirements:** This plan adheres to [List relevant guidelines and standards, e.g., FDA, ISO].

2. Objectives

- Ensure the warehouse maintains temperature between [range] and humidity between [range].

- Verify that all equipment operates as required to maintain these conditions.

3. Scope

- The validation will cover [Specify areas, e.g., storage rooms, loading docks].

- Equipment involved includes [List equipment, e.g., HVAC systems, temperature and humidity sensors, data loggers].

4. Responsibilities

- **Project Manager:** Oversees the entire validation process.

- **Validation Team:** Executes the validation activities and prepares documentation.

- **Quality Assurance:** Reviews and approves all validation documents.
- **Facility Team:** Ensures proper installation and maintenance of equipment.

5. Acceptance Criteria

- Temperature must be maintained between [range] °C.
- Humidity must be maintained between [range] %.
- All equipment must function without failure during the validation period.

6. Validation Activities

- **Installation Qualification (IQ):**
  - **Objective:** Verify correct installation of equipment.
  - **Activities:**
    - Check installation against manufacturer specifications.
    - Verify calibration of sensors.
    - Document installation process.

- **Operational Qualification (OQ):**
  - **Objective:** Ensure equipment operates within specified parameters.
  - **Activities:**
    - Conduct operational tests under normal conditions.
    - Test under worst-case scenarios.
    - Record test results and compare against acceptance criteria.

- **Performance Qualification (PQ):**
  - **Objective:** Validate maintenance of temperature and humidity over time.
  - **Activities:**
    - Monitor temperature and humidity for [specified period].
    - Collect and analyze data.
    - Document any deviations and corrective actions.

7. Schedule

- **Preparation:** [Start Date] to [End Date]
- **Installation Qualification:** [Start Date] to [End Date]
- **Operational Qualification:** [Start Date] to [End Date]
- **Performance Qualification:** [Start Date] to [End Date]

- **Review and Approval:** [Start Date] to [End Date]

8. Documentation

- **Installation Qualification Protocol and Report**
- **Operational Qualification Protocol and Report**
- **Performance Qualification Protocol and Report**
- **Validation Summary Report**

9. Approval

- **Prepared by:**
  - Name: _____
  - Title: _____
  - Signature: _____
  - Date: _____

- **Reviewed by:**
  - Name: _____
  - Title: _____
  - Signature: _____
  - Date: _____

- **Approved by:**
  - Name: _____
  - Title: _____
  - Signature: _____
  - Date: _____

This validation plan ensures a comprehensive and systematic approach to validating temperature and humidity control in the warehouse, maintaining product quality, and meeting regulatory requirements.

Step-by-Step Installation Qualification (IQ)

1. Preparation

- **Identify Equipment:** List all equipment that needs to be installed and qualified.

- **Gather Documentation:** Collect installation manuals, schematics, and manufacturer specifications for each piece of equipment.

- **Develop IQ Protocol:** Create an IQ protocol that outlines the steps and checks required to verify the installation.

2. Protocol Review and Approval

- **Draft IQ Protocol:** Draft the IQ protocol document.

- **Review:** Have the protocol reviewed by relevant personnel (e.g., quality assurance, facility team).

- **Approval:** Obtain approval from responsible parties before starting the IQ process.

3. Installation Verification

- **Verify Installation:** Ensure that all equipment is installed according to manufacturer specifications.

- **Check Locations:** Verify that all sensors, data loggers, and HVAC systems are installed in the correct locations.

- **Electrical Connections:** Ensure all electrical connections are secure and meet safety standards.

4. Calibration Verification

- **Calibration Certificates:** Check that all sensors and data loggers have current calibration certificates.

- **Perform Calibration:** If necessary, perform calibration of sensors and data loggers according to standard operating procedures.

- **Document Calibration:** Record the calibration data and attach certificates to the IQ report.

5. Documentation and Recording

- **Installation Checklist:** Use an installation checklist to verify each step of the installation process.

- **Photographic Evidence:** Take photographs of the installation for documentation purposes.

- **Installation Report:** Prepare an installation report detailing the installation process and results of verification checks.

6. Approval of IQ Report

- **Review IQ Report:** Have the completed IQ report reviewed by quality assurance and other relevant personnel.

- **Approval:** Obtain signatures from responsible parties to approve the IQ report.

Installation Qualification (IQ) Template

Installation Qualification (IQ) for Temperature and Humidity Control in Warehouse

1. Introduction

- **Objective:** To verify that all equipment for temperature and humidity control is installed correctly and operates as specified.

- **Scope:** This IQ covers the installation of HVAC systems, temperature sensors, humidity sensors, and data loggers in the warehouse.

2. Equipment Identification

- **HVAC System:** [Description, Model and Serial Number]

- **Temperature Sensors:** [Description, Model and Serial Number]

- **Humidity Sensors:** [Description, Model and Serial Number]

- **Data Loggers:** [Description, Model and Serial Number]

3. Documentation

- **Installation Manuals:** Attach relevant manuals and schematics.

- **Manufacturer Specifications:** Attach specifications for all equipment.

4. Installation Verification

- **Installation Checklist:**

Item	Specification	Verified (Yes/No)	Comments
HVAC System	Installed as per manual	Yes	
Temperature Sensors	Correct locations	Yes	
Humidity Sensors	Correct locations	Yes	
Data Loggers	Correct installation and positioning	Yes	
Electrical Connections	Secure and meet safety standards	Yes	

5. Calibration Verification

- **Calibration Records:**

Sensor/Logger	Calibration Date	Certificate Attached (Yes/No)	Calibration Due Date
Temperature Sensor 1			
Temperature Sensor 2			
Humidity Sensor 1			
Humidity Sensor 2			
Data Logger 1			
Data Logger 2			

6. Installation Report

- **Summary of Installation:** Provide a summary of the installation process.

- **Photographic Evidence:** Attach photographs of installed equipment.

- **Verification Results:** Summarize the results of installation verification checks.

7. Deviations and Resolutions

- **Deviations:** Document any deviations from the installation protocol and how they were resolved.

- **Corrective Actions:** Describe any corrective actions taken.

8. Approval

- **Prepared by:**

    o Name: _____

    o Title: _____

    o Signature: _____

    o Date: _____

- **Reviewed by:**

    o Name: _____

    o Title: _____

    o Signature: _____

    o Date: _____

- **Approved by:**

    o Name: _____

    o Title: _____

    o Signature: _____

    o Date: _____

This template provides a structured approach to documenting and verifying the installation of temperature and humidity control equipment in a warehouse, ensuring that all necessary steps are completed and properly documented.

Step-by-Step Operational Qualification (OQ)

1. Preparation

- **Identify Equipment and Systems:** List all equipment and systems to be tested during OQ.
- **Gather Documentation:** Collect operation manuals, standard operating procedures (SOPs), and manufacturer specifications.
- **Develop OQ Protocol:** Create an OQ protocol that outlines the operational tests and acceptance criteria.

2. Protocol Review and Approval

- **Draft OQ Protocol:** Draft the OQ protocol document.
- **Review:** Have the protocol reviewed by relevant personnel (e.g., quality assurance, facility team).
- **Approval:** Obtain approval from responsible parties before starting the OQ process.

3. Operational Testing

- **Normal Operating Conditions:** Test the equipment under normal operating conditions.
- **Worst-Case Scenarios:** Test the equipment under worst-case scenarios to ensure it can handle extreme conditions.
- **Functionality Checks:** Verify that each component operates as intended (e.g., HVAC system maintaining temperature, sensors providing accurate readings).
- **System Alarms and Fail-safes:** Test alarm systems and fail-safes to ensure they trigger correctly in case of deviations.

4. Data Collection and Recording

- **Record Test Results:** Document the results of each operational test, including data from sensors and loggers.
- **Data Analysis:** Analyze the collected data to ensure it meets the acceptance criteria.
- **Deviations and Resolutions:** Record any deviations from expected results and the corrective actions taken.

5. Documentation and Reporting

- **Operational Qualification Report:** Prepare an OQ report summarizing the testing process, results, and any deviations.
- **Attach Supporting Documents:** Include data logs, charts, and calibration certificates.

6. Approval of OQ Report

- **Review OQ Report:** Have the completed OQ report reviewed by quality assurance and other relevant personnel.
- **Approval:** Obtain signatures from responsible parties to approve the OQ report.

Example Operational Qualification (OQ) Template

Operational Qualification (OQ) for Temperature and Humidity Control in Warehouse

1. Introduction

- **Objective:** To verify that all equipment for temperature and humidity control operates correctly under normal and worst-case conditions.

- **Scope:** This OQ covers the operational testing of HVAC systems, temperature sensors, humidity sensors, and data loggers in the warehouse.

2. Equipment Identification

- **HVAC System:** [Description, Model and Serial Number]

- **Temperature Sensors:** [Description, Model and Serial Number]

- **Humidity Sensors:** [Description, Model and Serial Number]

- **Data Loggers:** [Description, Model and Serial Number]

3. Documentation

- **Operation Manuals:** Attach relevant manuals and SOPs.

- **Manufacturer Specifications:** Attach specifications for all equipment.

4. Operational Testing

Normal Operating Conditions:

- **Test Plan:**

Test Item	Description	Acceptance Criteria	Results Pass/Fail	Comments
HVAC System	Operate for 24 hours under normal conditions	Maintain temperature within [range]		
Temperature Sensors	Measure and record temperature every hour	Accuracy within ± [value] °C		
Humidity Sensors	Measure and record humidity every hour	Accuracy within ± [value] %		
Data Loggers	Log data continuously for 24 hours	Accurate data logging without gaps		

Worst-Case Scenarios:

- **Test Plan:**

Test Item	Description	Acceptance Criteria	Results (Pass/Fail)	Comments
HVAC System	Operate under maximum load for 8 hours	Maintain temperature within [range]		
Temperature Sensors	Measure temperature during high fluctuations	Accuracy within ± [value] °C		
Humidity Sensors	Measure humidity during high fluctuations	Accuracy within ± [value] %		
Data Loggers	Log data under fluctuating conditions	Accurate data logging without errors		

System Alarms and Fail-safes:

- **Test Plan:**

Test Item	Description	Acceptance Criteria	Results (Pass/Fail)	Comments
HVAC Alarm	Simulate HVAC failure	Alarm triggers within [time] seconds		
Sensor Fail-safe	Simulate sensor failure	System switches to backup sensor		
Data Logger Alarm	Simulate data logger error	Alarm triggers and logs error		

5. Data Collection and Recording

- **Test Results:** Document the results of each test in the provided tables.

- **Data Logs:** Attach data logs, charts, and calibration certificates.

- **Analysis:** Analyze data to ensure compliance with acceptance criteria.

- **Deviations:** Record any deviations and corrective actions.

6. Operational Qualification Report

- **Summary of Testing:** Provide a summary of the operational tests conducted.

- **Results:** Summarize the results of the tests, highlighting any deviations and resolutions.

- **Conclusion:** State whether the equipment operates within the specified parameters.

7. Deviations and Resolutions

- **Deviations:** Document any deviations from the operational tests.

- **Corrective Actions:** Describe any corrective actions taken to address deviations.

8. Approval

- **Prepared by:**
  - Name: _____
  - Title: _____
  - Signature: _____
  - Date: _____

- **Reviewed by:**
  - Name: _____
  - Title: _____
  - Signature: _____
  - Date: _____

- **Approved by:**
  - Name: _____
  - Title: _____
  - Signature: _____
  - Date: _____

Step-by-Step Performance Qualification (PQ)

1. Preparation

- **Identify Equipment and Systems:** List all equipment and systems to be monitored during PQ.
- **Define Acceptance Criteria:** Establish the acceptable range for temperature and humidity levels.
- **Develop PQ Protocol:** Create a PQ protocol that outlines the monitoring procedures and acceptance criteria.

2. Protocol Review and Approval

- **Draft PQ Protocol:** Draft the PQ protocol document.
- **Review:** Have the protocol reviewed by relevant personnel (e.g., quality assurance, facility team).
- **Approval:** Obtain approval from responsible parties before starting the PQ process.

3. Monitoring Plan

- **Duration:** Define the duration of the monitoring period (e.g., 30 days).
- **Frequency:** Specify the frequency of data collection (e.g., every 15 minutes).
- **Data Loggers Placement:** Place data loggers in critical locations within the warehouse.

4. Performance Testing

- **Continuous Monitoring:** Monitor temperature and humidity continuously throughout the defined period.
- **Data Collection:** Collect data at specified intervals using calibrated data loggers.
- **Environmental Variations:** Ensure testing covers normal and extreme weather conditions to validate performance under all possible scenarios.

5. Data Analysis

- **Compile Data:** Gather all data collected during the monitoring period.
- **Analyze Data:** Compare the collected data against the defined acceptance criteria.
- **Identify Deviations:** Note any deviations from the acceptable range and analyze their causes.

6. Documentation and Reporting

- **Performance Qualification Report:** Prepare a PQ report summarizing the monitoring process, results, and any deviations.
- **Attach Supporting Documents:** Include data logs, charts, and calibration certificates.

7. Approval of PQ Report

- **Review PQ Report:** Have the completed PQ report reviewed by quality assurance and other relevant personnel.
- **Approval:** Obtain signatures from responsible parties to approve the PQ report.

Example Performance Qualification (PQ) Template

Performance Qualification (PQ) for Temperature and Humidity Control in Warehouse

1. Introduction

- **Objective:** To verify that the warehouse maintains temperature and humidity within specified ranges over an extended period.

- **Scope:** This PQ covers the performance monitoring of HVAC systems, temperature sensors, humidity sensors, and data loggers in the warehouse.

2. Equipment Identification

- **HVAC System:** [Description, Model and Serial Number]

- **Temperature Sensors:** [Description, Model and Serial Number]

- **Humidity Sensors:** [Description, Model and Serial Number]

- **Data Loggers:** [Description, Model and Serial Number]

3. Documentation

- **Operation Manuals:** Attach relevant manuals and SOPs.

- **Manufacturer Specifications:** Attach specifications for all equipment.

4. Monitoring Plan

- **Duration:** 30 days

- **Frequency:** Every 15 minutes

- **Data Loggers Placement:**

  o Logger 1: North end of the warehouse

  o Logger 2: South end of the warehouse

  o Logger 3: Center of the warehouse

5. Performance Testing

Monitoring Data:

- **Temperature Monitoring:**

Time Interval	Logger 1 (°C)	Logger 2 (°C)	Logger 3 (°C)	Acceptance Criteria (°C)	Pass/Fail	Comments
Day 1 - 00:00						

Time Interval	Logger 1 (°C)	Logger 2 (°C)	Logger 3 (°C)	Acceptance Criteria (°C)	Pass/Fail	Comments
Day 1 - 00:15						
...						
Day 30 - 23:45						

- **Humidity Monitoring:**

Time Interval	Logger 1 (%)	Logger 2 (%)	Logger 3 (%)	Acceptance Criteria (%)	Pass/Fail	Comments
Day 1 - 00:00						
Day 1 - 00:15						
...						
Day 30 - 23:45						

6. Data Analysis

- **Compile Data:** Summarize the data collected from each logger.
- **Compare Data:** Compare the data against the acceptance criteria.
- **Analyze Deviations:** Document any deviations and their causes.

7. Performance Qualification Report

- **Summary of Monitoring:** Provide a summary of the monitoring process.
- **Results:** Summarize the results, highlighting any deviations and resolutions.
- **Conclusion:** State whether the warehouse maintains the required temperature and humidity ranges.

8. Deviations and Resolutions

- **Deviations:** Document any deviations from the performance criteria.
- **Corrective Actions:** Describe any corrective actions taken to address deviations.

9. Approval

- **Prepared by:**
  - Name: _____
  - Title: _____
  - Signature: _____
  - Date: _____

- **Reviewed by:**
  - Name: _____
  - Title: _____
  - Signature: _____
  - Date: _____

- **Approved by:**
  - Name: _____
  - Title: _____
  - Signature: _____
  - Date: _____

This template provides a structured approach to documenting and verifying the performance of temperature and humidity control equipment in a warehouse, ensuring that all necessary tests are completed and properly documented.

Step-by-Step Risk Assessment

1. Preparation

- **Assemble a Team:** Gather a team of experts from relevant departments such as quality assurance, warehouse management, facility, and safety.

- **Define Objectives:** Clearly outline the objectives of the risk assessment.

- **Collect Information:** Gather relevant information about the warehouse, equipment, processes, and environmental conditions.

2. Identify Risks

- **Brainstorm Potential Risks:** Conduct brainstorming sessions with the team to identify potential risks related to temperature and humidity control.

- **Categorize Risks:** Group the identified risks into categories such as equipment failure, environmental factors, human error, and operational issues.

3. Evaluate Risks

- **Assess Impact:** Determine the potential impact of each risk on warehouse operations, product quality, and regulatory compliance.

- **Assess Likelihood:** Estimate the likelihood of each risk occurring based on historical data, expert judgment, and environmental conditions.

- **Risk Matrix:** Use a risk matrix to prioritize risks based on their impact and likelihood.

4. Mitigation Strategies

- **Develop Mitigation Plans:** For each identified risk, develop strategies to mitigate or reduce the impact and likelihood of the risk.

- **Implement Controls:** Identify and implement control measures such as equipment maintenance schedules, training programs, and environmental monitoring systems.

5. Document and Review

- **Prepare Risk Assessment Report:** Document the risk assessment process, including identified risks, evaluations, and mitigation strategies.

- **Review and Approve:** Have the risk assessment report reviewed and approved by relevant stakeholders.

6. Monitor and Update

- **Continuous Monitoring:** Continuously monitor the effectiveness of mitigation strategies and control measures.

- **Regular Updates:** Regularly update the risk assessment to reflect changes in processes, equipment, or environmental conditions.

Example Risk Assessment Template

Risk Assessment for Temperature and Humidity Control in Warehouse

1. Introduction

- **Objective:** To identify, evaluate, and mitigate risks associated with temperature and humidity control in the warehouse.

- **Scope:** This risk assessment covers all temperature and humidity control systems and processes in the warehouse.

2. Risk Identification

Potential Risks:

- **Equipment Failure:** Failure of HVAC systems, temperature sensors, or humidity sensors.

- **Environmental Factors:** Extreme weather conditions affecting internal temperature and humidity.

- **Human Error:** Incorrect settings or failure to monitor equipment.

- **Operational Issues:** Power outages, insufficient maintenance, and calibration issues.

3. Risk Evaluation

Risk Matrix:

Risk Category	Risk Description	Impact (High/Medium/Low)	Likelihood (High/Medium/Low)	Priority (High/Medium/Low)
Equipment Failure	HVAC system failure	High	Medium	High
Environmental	Extreme weather conditions	Medium	High	High
Human Error	Incorrect equipment settings	Medium	Medium	Medium
Operational Issues	Power outage	High	Low	Medium

4. Mitigation Strategies

Mitigation Plans:

Risk Category	Risk Description	Mitigation Strategy	Responsible Party
Equipment Failure	HVAC system failure	Regular maintenance and servicing; install backup systems	Facility Team

Risk Category	Risk Description	Mitigation Strategy	Responsible Party
Environmental	Extreme weather conditions	Insulate warehouse; enhance HVAC capacity	Warehouse Management
Human Error	Incorrect equipment settings	Regular training and SOPs	Quality Assurance
Operational Issues	Power outage	Install backup generators and for critical equipment	Facility Team

5. Documentation and Review

Risk Assessment Report:

- **Summary of Identified Risks:** Provide a summary of all identified risks.

- **Evaluation Results:** Summarize the evaluation of risks using the risk matrix.

- **Mitigation Strategies:** Detail the mitigation strategies and their implementation status.

- **Approval:**

    o **Prepared by:**

        - Name: _____

        - Title: _____

        - Signature: _____

        - Date: _____

    o **Reviewed by:**

        - Name: _____

        - Title: _____

        - Signature: _____

        - Date: _____

    o **Approved by:**

        - Name: _____

        - Title: _____

        - Signature: _____

        - Date: _____

6. Monitoring and Update

Continuous Monitoring:

- **Effectiveness Monitoring:** Regularly monitor the effectiveness of implemented mitigation strategies.

- **Review Schedule:** Update the risk assessment [e.g., annually, bi-annually] or when significant changes occur.

This template provides a structured approach to identifying, evaluating, and mitigating risks associated with temperature and humidity control in a warehouse, ensuring that all necessary steps are completed and properly documented.

Step-by-Step Data Analysis

1. Preparation

- **Define Objectives:** Clearly outline the objectives of the data analysis, such as ensuring temperature and humidity levels are within specified ranges.

- **Collect Data:** Gather all relevant data from temperature and humidity sensors, data loggers, and monitoring systems over the specified period.

- **Organize Data:** Compile the data in a structured format, such as a spreadsheet or database, for easy analysis.

2. Data Cleaning

- **Check for Completeness:** Ensure all expected data points are present.

- **Identify Outliers:** Look for any data points that are significantly different from others and investigate their causes.

- **Correct Errors:** Fix any identified errors or anomalies in the data, such as incorrect readings or timestamps.

3. Descriptive Statistics

- **Calculate Summary Statistics:** Compute mean, median, mode, standard deviation, and range for temperature and humidity data.

- **Visualize Data:** Create charts and graphs (e.g., line graphs, histograms) to visualize data trends and distributions.

4. Trend Analysis

- **Identify Patterns:** Analyze the data for any recurring patterns or trends over time.

- **Seasonal Effects:** Consider any seasonal variations that might affect temperature and humidity levels.

- **Correlation Analysis:** Check for correlations between temperature and humidity, and other factors like time of day or external weather conditions.

5. Compliance Check

- **Compare Against Criteria:** Compare the collected data against the defined acceptance criteria or regulatory standards.

- **Identify Deviations:** Note any instances where the data falls outside the acceptable range.

- **Document Deviations:** Record details of any deviations, including the magnitude and duration.

6. Root Cause Analysis

- **Investigate Deviations:** Perform root cause analysis for any deviations to understand why they occurred.

- **Identify Contributing Factors:** Determine if equipment failure, environmental factors, or operational issues contributed to the deviations.

7. Recommendations

- **Suggest Improvements:** Based on the analysis, suggest improvements to equipment, processes, or controls to maintain temperature and humidity within acceptable ranges.

- **Develop Action Plan:** Create an action plan to implement the suggested improvements.

8. Documentation and Reporting

- **Prepare Data Analysis Report:** Summarize the analysis process, results, deviations, root causes, and recommendations.

- **Review and Approve:** Have the report reviewed and approved by relevant stakeholders.

Example Data Analysis Template

Data Analysis for Temperature and Humidity Control in Warehouse

1. Introduction

- **Objective:** To analyze temperature and humidity data to ensure they are within specified ranges.

- **Scope:** This analysis covers data collected from temperature and humidity sensors in the warehouse over [time period].

2. Data Collection

- **Data Sources:** List the sensors and data loggers used.

- **Time Period:** Specify the period during which data was collected.

3. Data Cleaning

- **Completeness Check:**

Date	Time	Logger ID	Temperature (°C)	Humidity (%)	Status
2024-05-01	00:00	Logger 1	22.5	50.2	Complete
2024-05-01	00:15	Logger 1	22.7	50.0	Complete
...	...	...	...	...	...
2024-05-30	23:45	Logger 3	23.1	49.8	Complete

- **Outlier Identification:**

Date	Time	Logger ID	Temperature (°C)	Humidity (%)	Outlier (Y/N)
2024-05-15	12:00	Logger 2	30.0	60.5	Y
2024-05-22	14:30	Logger 3	20.1	40.3	Y

Date	Time	Logger ID	Temperature (°C)	Humidity (%)	Outlier (Y/N)
...	...	...	...	...	...

4. Descriptive Statistics

Temperature:

Statistic	Value
Mean	22.8 °C
Median	22.7 °C
Mode	22.5 °C
Standard Deviation	1.2 °C
Range	20.1-30.0 °C

Humidity:

Statistic	Value
Mean	50.5 %
Median	50.3 %
Mode	50.0 %
Standard Deviation	2.5 %
Range	40.3-60.5 %

Charts and Graphs:

- Include line graphs showing temperature and humidity trends over time.

- Include histograms showing the distribution of temperature and humidity values.

5. Trend Analysis

Temperature Trends:

- Identify daily and weekly patterns.

- Highlight any periods of significant temperature changes.

Humidity Trends:

- Identify daily and weekly patterns.

- Highlight any periods of significant humidity changes.

6. Compliance Check

Temperature Compliance:

Time Period	Logger ID	Temperature Range (°C)	Acceptance Criteria (°C)	Pass/Fail	Comments
2024-05-01 to 2024-05-30	Logger 1	20.1-30.0	20.0-25.0	Fail	Deviation on 2024-05-15

Humidity Compliance:

Time Period	Logger ID	Humidity Range (%)	Acceptance Criteria (%)	Pass/Fail	Comments
2024-05-01 to 2024-05-30	Logger 2	40.3-60.5	45.0-55.0	Fail	Deviation on 2024-05-22

7. Root Cause Analysis

- **Deviation on 2024-05-15:**

  - **Possible Cause:** HVAC system failure.

  - **Investigation:** Check maintenance logs and sensor calibration.

  - **Resolution:** Repair HVAC system, recalibrate sensor.

- **Deviation on 2024-05-22:**

  - **Possible Cause:** Extreme external weather conditions.

  - **Investigation:** Review external weather data and warehouse insulation.

  - **Resolution:** Enhance insulation and review HVAC capacity.

8. Recommendations

- **Equipment Improvements:** Upgrade HVAC system for better temperature control.

- **Process Improvements:** Implement regular sensor calibration and maintenance schedule.

- **Environmental Controls:** Enhance warehouse insulation to mitigate external weather impact.

9. Documentation and Reporting

Data Analysis Report:

- **Summary of Analysis:** Provide a summary of the data analysis process.

- **Results:** Summarize the findings, including compliance check results and deviations.

- **Recommendations:** Detail the recommended improvements and action plan.

Approval:

Prepared by:

- Name: _____

- Title: _____

- Signature: _____

- Date: _____

Reviewed by:

- Name: _____

- Title: _____

- Signature: _____

- Date: _____

Approved by:

- Name: _____

- Title: _____

- Signature: _____

- Date: _____

This template provides a structured approach to analyzing temperature and humidity data in a warehouse, ensuring that all necessary steps are completed and properly documented.

Step-by-Step Determination for Revalidation

1. Define Revalidation Triggers

- **Regulatory Requirements:** Identify revalidation intervals specified by regulatory bodies or industry standards.

- **Equipment Changes:** Determine revalidation needs when there are changes to equipment or systems (e.g., installation of new HVAC units, sensor upgrades).

- **Process Changes:** Consider revalidation when there are significant changes to warehouse processes that could affect temperature and humidity control.

- **Environmental Changes:** Account for revalidation if there are substantial changes to the warehouse environment, such as structural modifications or changes in insulation.

- **Deviations and Failures:** Identify the need for revalidation following significant deviations, equipment failures, or out-of-specification results.

- **Scheduled Intervals:** Establish regular revalidation intervals based on risk assessment and historical performance data (e.g., annually, bi-annually).

2. Develop a Revalidation Plan

- **Documentation:** Create a revalidation plan that includes all identified triggers, timelines, and responsible parties.

- **Approval:** Have the revalidation plan reviewed and approved by relevant stakeholders, such as quality assurance and warehouse management.

3. Monitoring and Recording Triggers

- **Continuous Monitoring:** Implement a system for continuous monitoring of temperature and humidity to detect deviations.

- **Record Keeping:** Maintain accurate records of equipment changes, process modifications, and any deviations or failures.

- **Periodic Reviews:** Conduct periodic reviews to assess if any triggers for revalidation have occurred.

4. Initiate Revalidation Process

- **Trigger Identification:** When a revalidation trigger is identified, initiate the revalidation process.

- **Plan Execution:** Follow the revalidation plan, conducting necessary validation activities such as Installation Qualification (IQ), Operational Qualification (OQ), and Performance Qualification (PQ).

5. Document and Approve Revalidation

- **Revalidation Report:** Prepare a revalidation report summarizing the process, findings, and any corrective actions taken.

- **Review and Approval:** Have the revalidation report reviewed and approved by relevant stakeholders.

Example Revalidation Trigger Template

Revalidation Trigger Plan for Temperature and Humidity Control in Warehouse

1. Introduction

- **Objective:** To ensure continued compliance and optimal performance of temperature and humidity control systems in the warehouse.

- **Scope:** This plan outlines the triggers and procedures for revalidation.

2. Revalidation Triggers

Regulatory Requirements:

- Revalidation every [e.g., 2 years] as per [regulatory standard].

Equipment Changes:

- Installation of new HVAC units.

- Upgrade or replacement of temperature and humidity sensors.

- Major repairs or modifications to existing equipment.

Process Changes:

- Changes to warehouse storage practices that could affect environmental conditions.

- Implementation of new procedures for handling temperature-sensitive products.

Environmental Changes:

- Structural modifications to the warehouse (e.g., new insulation, changes to the building layout).

- Changes in external weather conditions that impact internal temperature and humidity.

Deviations and Failures:

- Significant deviations from temperature and humidity specifications.

- Equipment failures or breakdowns affecting environmental control.

Scheduled Intervals:

- Regular revalidation every [e.g., 1 year] based on historical data and risk assessment.

3. Monitoring and Recording Triggers

- **Continuous Monitoring:**

  - Implement continuous monitoring of temperature and humidity levels.

  - Use automated data logging systems to detect and record deviations.

- **Record Keeping:**

  - Maintain logs of all equipment changes, process modifications, and deviations.

- o Document any corrective actions taken in response to deviations or failures.

- **Periodic Reviews:**

  - o Conduct monthly or quarterly reviews of monitoring data and records to assess the need for revalidation.

4. Initiate Revalidation Process

- **Trigger Identification:**

  - o When a trigger is identified, notify relevant stakeholders and initiate the revalidation process.

- **Plan Execution:**

  - o Follow the revalidation plan, conducting IQ, OQ, and PQ as needed.

  - o Document all revalidation activities and findings.

5. Documentation and Approval

- **Revalidation Report:**

  - o Prepare a comprehensive report summarizing the revalidation process, findings, and any corrective actions taken.

- **Review and Approval:**

  - o Have the revalidation report reviewed by quality assurance and warehouse management.

  - o Obtain approval from responsible parties to confirm successful revalidation.

6. Approval

- **Prepared by:**

  - o Name: _____

  - o Title: _____

  - o Signature: _____

  - o Date: _____

- **Reviewed by:**

  - o Name: _____

  - o Title: _____

  - o Signature: _____

  - o Date: _____

- **Approved by:**

  - o Name: _____

- o Title: _____
- o Signature: _____
- o Date: _____

This template provides a structured approach to identifying and managing revalidation triggers, ensuring that temperature and humidity control systems in the warehouse remain effective and compliant with regulatory standards.

Step-by-Step Guide to Resolving Deviations During Validation

When deviations occur during the validation of temperature and humidity control systems in a warehouse, it's crucial to follow a structured approach to identify, analyze, and correct the deviations. Here is a step-by-step guide to resolving deviations during validation, along with an example to illustrate the process.

1. Detection of Deviation

- **Monitor Validation Data:** Continuously monitor data collected during validation.

- **Identify Deviations:** Identify any data points or trends that fall outside of the predefined specifications or acceptance criteria.

2. Document the Deviation

- **Record Details:** Record details of the deviation, including date, time, specific measurements, and conditions at the time of deviation.

- **Deviation Report:** Create a deviation report documenting the specifics of the deviation.

3. Initial Assessment

- **Impact Assessment:** Conduct an initial assessment to determine the impact of the deviation on the validation process and overall system performance.

- **Immediate Actions:** Implement immediate corrective actions if necessary to mitigate any adverse effects.

4. Root Cause Analysis

- **Form an Investigation Team:** Assemble a team that includes quality assurance, engineering, and relevant stakeholders.

- **Collect Data:** Gather all relevant data, including environmental conditions, equipment logs, and operational procedures.

- **Analyze Data:** Use root cause analysis tools such as the 5 Whys, Fishbone Diagram, or Fault Tree Analysis to identify the underlying cause(s) of the deviation.

5. Develop Corrective Actions

- **Action Plan:** Develop a corrective action plan to address the root cause of the deviation.

- **Implement Corrective Actions:** Execute the corrective actions as per the plan.

- **Document Actions:** Record all corrective actions taken, including dates, responsible parties, and outcomes.

6. Revalidate the System

- **Revalidation Plan:** Create a revalidation plan to ensure that the corrective actions have resolved the deviation and the system meets the acceptance criteria.

- **Execute Revalidation:** Perform the revalidation activities according to the plan, including repeating relevant tests and measurements.

- **Verify Results:** Verify that the revalidation results are within the acceptable range.

7. Preventive Actions

- **Update Procedures:** Update Standard Operating Procedures (SOPs) to include new preventive measures that address the root cause.

- **Training:** Conduct training sessions for staff to ensure they understand the updated procedures and preventive measures.

8. Documentation and Approval

- **Final Report:** Prepare a final report summarizing the deviation, root cause analysis, corrective actions, revalidation results, and preventive measures.

- **Review and Approval:** Have the report reviewed and approved by relevant stakeholders.

Example: Resolving a Deviation During Validation

1. Detection of Deviation

- **Scenario:** During the Operational Qualification (OQ) phase of validating the warehouse's HVAC system, a temperature reading of 28°C was recorded, exceeding the specified range of 20-25°C.

2. Document the Deviation

- **Deviation Report:**

    o **Date:** 2024-06-10

    o **Time:** 14:00

    o **Temperature Reading:** 28°C

    o **Specification Range:** 20-25°C

3. Initial Assessment

- **Impact Assessment:** The deviation could affect temperature-sensitive products stored in the warehouse. Immediate action is required to prevent potential product quality issues.

- **Immediate Actions:** Adjusted HVAC settings to bring temperature back within the specified range.

4. Root Cause Analysis

- **Investigation Team:** Quality Assurance Manager, HVAC Technician, Warehouse Supervisor.

- **Data Collected:** Review of HVAC system logs, temperature sensor calibration records, recent maintenance activities.

- **Analysis:** Using the Fishbone Diagram, the team identified a possible cause: a recently replaced temperature sensor was not properly calibrated.

5. Develop Corrective Actions

- **Action Plan:**

- o Recalibrate the temperature sensor.

- o Inspect and recalibrate all other temperature sensors.

- o Review and adjust HVAC settings to ensure stable temperature control.

- **Implement Corrective Actions:** Recalibrated the faulty sensor and inspected all other sensors.

- **Documentation:** Recorded calibration activities and adjustments made to the HVAC system.

6. Revalidate the System

- **Revalidation Plan:** Repeat the OQ phase for the temperature control system, focusing on temperature stability within the specified range.

- **Execute Revalidation:** Conducted temperature measurements over a 48-hour period.

- **Verify Results:** All temperature readings were within the 20-25°C range.

7. Preventive Actions

- **Update Procedures:** Revised the SOPs to include regular calibration checks for temperature sensors, especially after maintenance activities.

- **Training:** Conducted training for facility staff on proper calibration procedures and the importance of accurate sensor readings.

8. Documentation and Approval

- **Final Report:**

  - o **Summary of Deviation:** Temperature reading of 28°C during OQ phase.

  - o **Root Cause Analysis:** Identified faulty calibration of a replaced temperature sensor.

  - o **Corrective Actions:** Recalibrated sensors, adjusted HVAC settings.

  - o **Revalidation Results:** Successful revalidation with all temperatures within the specified range.

  - o **Preventive Actions:** Updated SOPs, conducted staff training.

- **Review and Approval:**

  - o **Prepared by:**

    - Name: John Lee

    - Title: Quality Assurance Manager

    - Signature: _____

    - Date: 2024-06-12

  - o **Reviewed by:**

    - Name: Jane Smith

- Title: Warehouse Supervisor
- Signature: _____
- Date: 2024-06-13

- **Approved by:**
  - Name: Emily Brown
  - Title: Compliance Officer
  - Signature: _____
  - Date: 2024-06-14

This template provides a comprehensive approach to resolving deviations during validation, ensuring that all necessary steps are completed and properly documented to maintain system integrity and compliance.

Step-by-Step Non-Conformance Investigation and Correction

1. Detection and Initial Response

- **Monitor Continuously:** Use automated data loggers and monitoring systems to continuously track temperature and humidity.

- **Identify Non-Conformance:** Detect deviations from the acceptable range for temperature and humidity.

- **Immediate Action:** Take immediate action to mitigate any potential damage (e.g., adjusting HVAC settings, moving sensitive products to a different location).

2. Record Non-Conformance

- **Document the Incident:** Record details of the non-conformance including date, time, location, and extent of deviation.

- **Initial Assessment:** Assess the immediate impact on stored products and processes.

3. Containment

- **Prevent Further Impact:** Implement measures to prevent further impact on products, such as using temporary cooling/heating units or humidity control devices.

- **Quarantine Affected Products:** Move affected products to a quarantine area until a full assessment is conducted.

4. Root Cause Analysis

- **Form Investigation Team:** Assemble a team including quality assurance, facility, and warehouse management.

- **Gather Data:** Collect data related to the incident, including environmental data, equipment logs, and maintenance records.

- **Identify Possible Causes:** Use tools such as the 5 Whys, Fishbone Diagram (Ishikawa), or Fault Tree Analysis to identify potential causes.

- **Analyze Data:** Evaluate the gathered data to pinpoint the root cause(s) of the deviation.

5. Corrective Actions

- **Develop Action Plan:** Create a plan to correct the root cause(s) identified.

- **Implement Corrections:** Execute corrective actions, such as repairing or recalibrating equipment, updating SOPs, or providing additional training to staff.

- **Document Actions:** Record all corrective actions taken, including the responsible parties and timelines.

6. Preventive Actions

- **Review Procedures:** Review existing procedures to identify any gaps that may have contributed to the non-conformance.

- **Update SOPs:** Update Standard Operating Procedures (SOPs) to include new preventive measures.

- **Training:** Conduct training sessions for staff to ensure they understand new procedures and preventive measures.

7. Verification and Validation

- **Verify Corrections:** Verify that the corrective actions have been implemented effectively and the issue is resolved.

- **Monitor:** Continue to monitor temperature and humidity levels to ensure they remain within specifications.

- **Revalidate:** If necessary, revalidate the affected systems to confirm they are functioning correctly.

8. Reporting and Approval

- **Prepare Report:** Document the entire process, including the initial detection, investigation findings, corrective and preventive actions, and verification results.

- **Review and Approval:** Have the report reviewed and approved by relevant stakeholders.

Example Non-Conformance Investigation and Correction Template

Non-Conformance Investigation and Correction Report

1. Incident Detection and Initial Response

- **Date and Time:** _____

- **Location:** _____

- **Description of Deviation:** _____

- **Immediate Actions Taken:** _____

2. Non-Conformance Record

- **Temperature Reading:** _____ °C (specification: _____ °C)

- **Humidity Reading:** _____ % (specification: _____ %)

- **Initial Impact Assessment:** _____

3. Containment Actions

- **Containment Measures Implemented:** _____

- **Products Quarantined:** Yes / No

- **Details of Quarantined Products:** _____

4. Root Cause Analysis

- **Investigation Team:** _____

- **Data Collected:** _____

- **Tools Used:** (e.g., 5 Whys, Fishbone Diagram) _____
- **Root Cause Identified:** _____

5. Corrective Actions

- **Action Plan:** _____
- **Corrective Actions Implemented:** _____
- **Responsible Parties:** _____
- **Completion Date:** _____

6. Preventive Actions

- **Procedures Reviewed:** _____
- **SOP Updates:** _____
- **Staff Training Conducted:** _____

7. Verification and Validation

- **Verification Methods:** _____
- **Results of Verification:** _____
- **Revalidation Required:** Yes / No
- **Revalidation Results:** _____

8. Reporting and Approval

- **Report Prepared by:**
  - Name: _____
  - Title: _____
  - Signature: _____
  - Date: _____
- **Reviewed by:**
  - Name: _____
  - Title: _____
  - Signature: _____
  - Date: _____
- **Approved by:**
  - Name: _____
  - Title: _____

- o Signature: _____

- o Date: _____

This template ensures a structured approach to investigating and correcting temperature and humidity non-conformances, maintaining product quality, and compliance with regulatory requirements.

Note